Published in the U.S.A. by Bara Publishing
BaraPublishing.com
For inquiries: info@barapublishing.com

© 2022 by Bethany & Ryan Bomberger

Book editing, design and layout by Ryan Scott Bomberger

🇺🇸 **Printed in the U.S.A.** 🇺🇸
Signature Book Printing
sbpbooks.com

ISBN: 978-0-9972036-3-9
Library of Congress Control Number - 2022920472

*Although we want you to share this wonderful book with everyone you know, we highly recommend gifting it, not copying it. Of course, if you want to promote how much you love the book on social media, feel free to post the cover or a favorite image in the book! Here comes the legal stuff. No part of this publication may be reproduced, distributed, or transmitted in any form or by any means, including photocopying, recording, or other electronic or mechanical methods (well except for the whole **tell-your-friends-and-family-online** caveat), without the prior written permission of both the copyright owner and the publisher of the book.*

DEDICATION

To all the incredible women in our lives, thank you for loving, nurturing, inspiring, and shaping us into who we are today.

To every girl out there...you are so precious. God designed you beautifully and uniquely. Whether you're a princess or whether you're a warrior (or if you're a princess warrior), *love* who you were created to be!

Guys, no matter what society says, you *always* matter. Don't let our culture erase women.

—Bethany & Ryan Bomberger

She is not he.

She is not we.

She is she.

Beautifully.

She's a doctor.

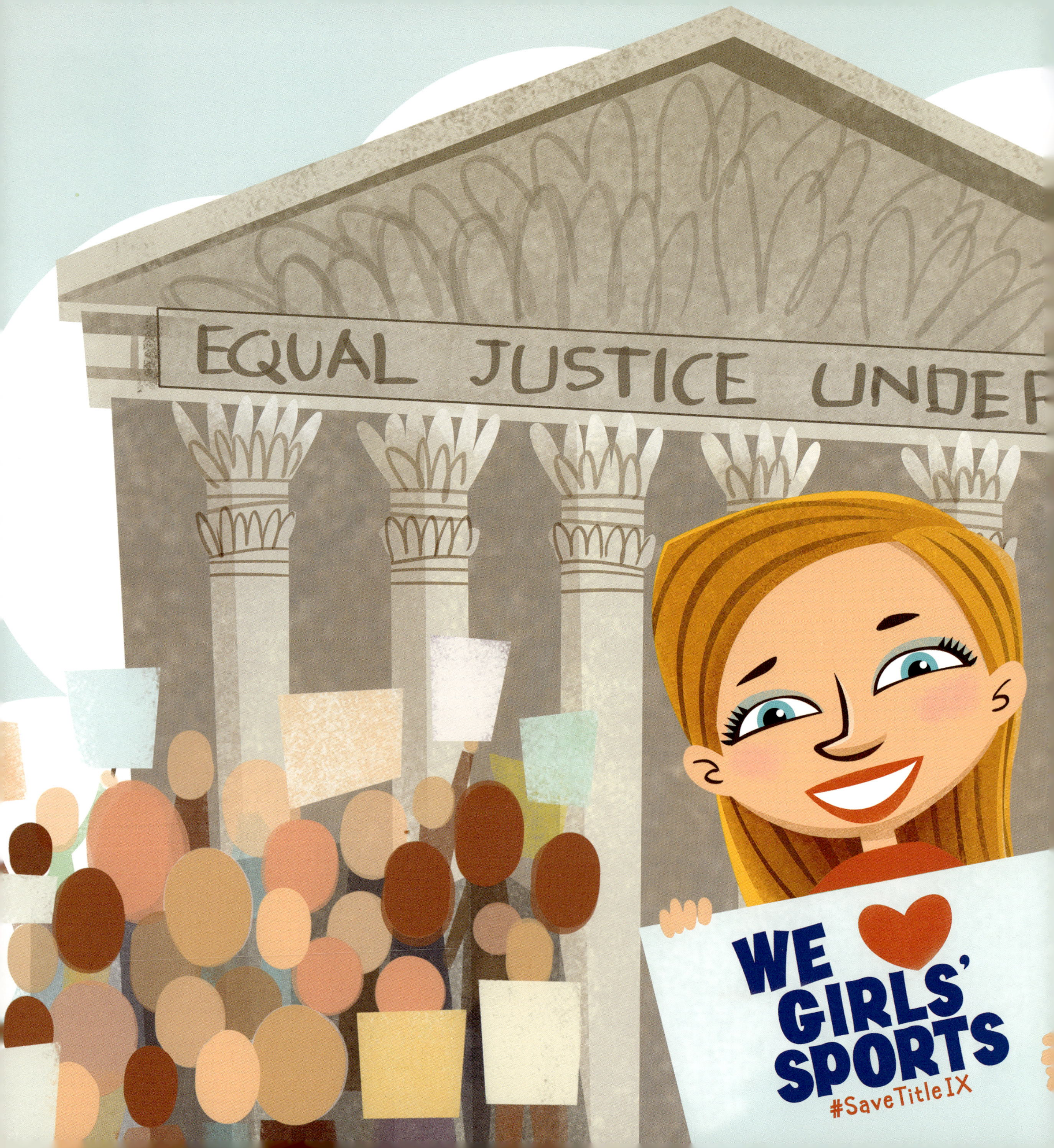
EQUAL JUSTICE UNDER
WE ♥ GIRLS' SPORTS
#SaveTitleIX

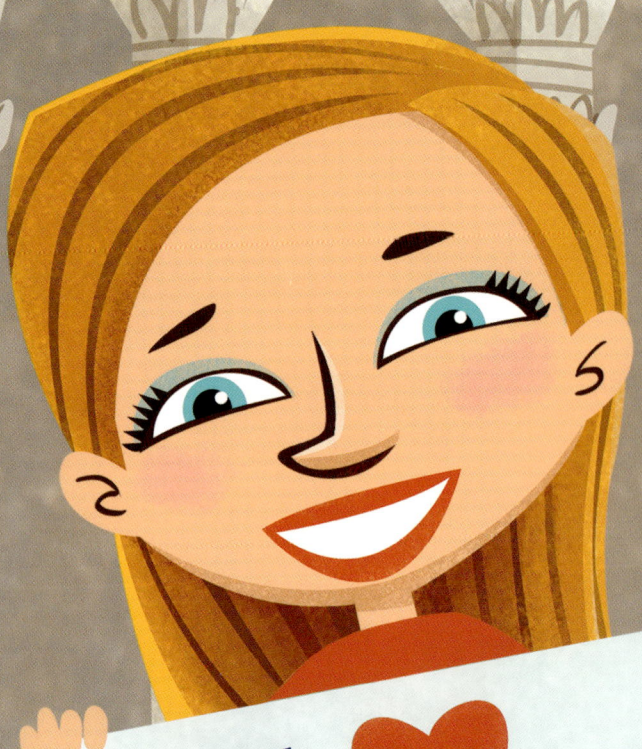

She's a justice-seeking lawyer.

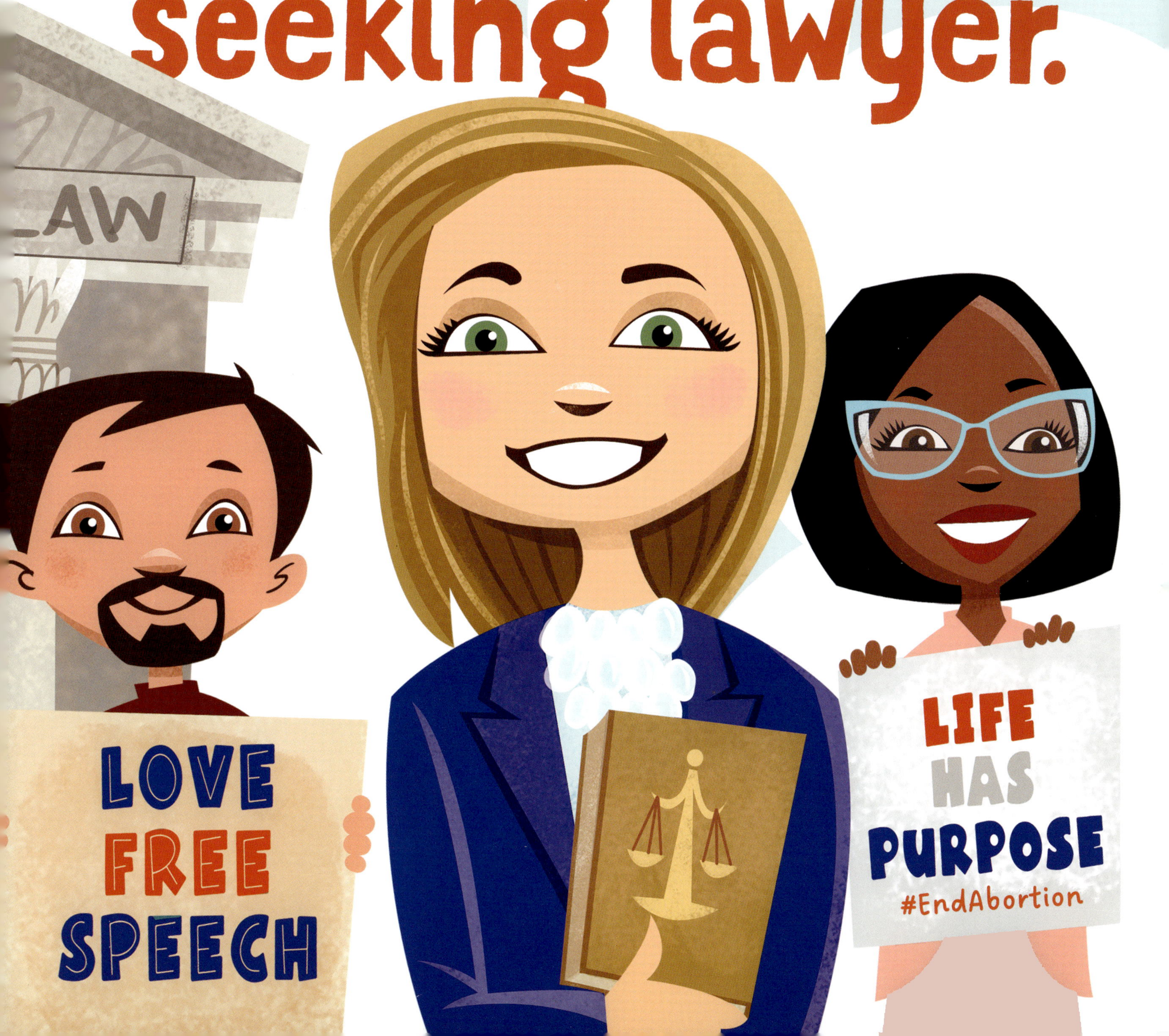

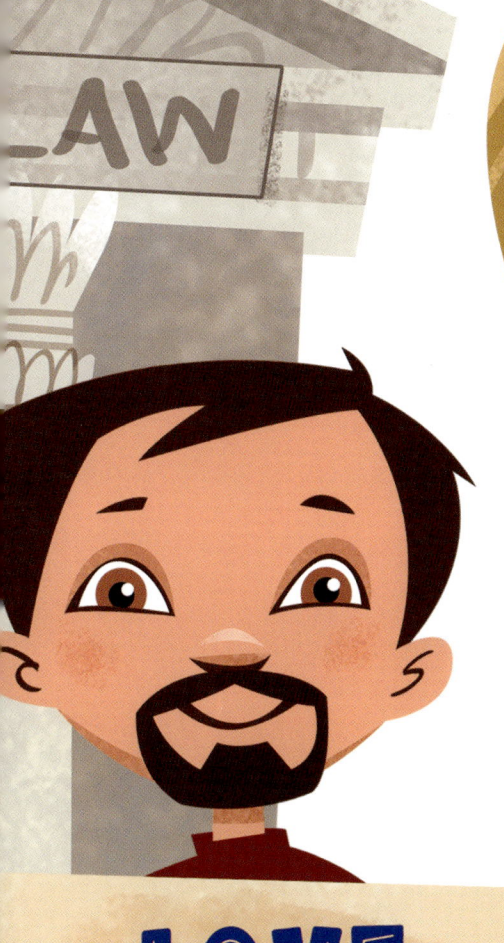

She's a leader.

She's a favorite grade school teacher.

When she is she

she is free

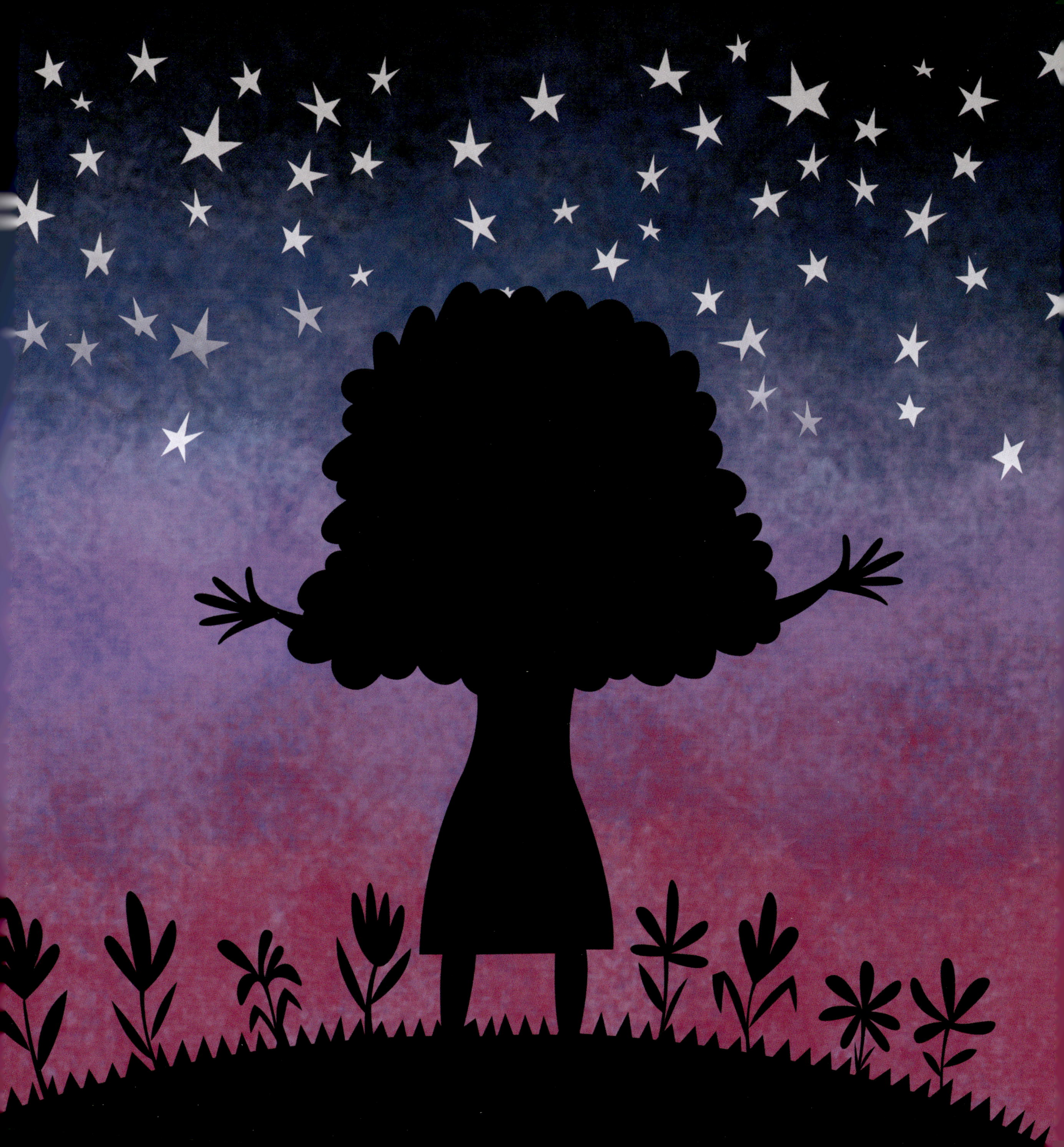

The

"READ IT AGAIN GRAMMY!"

End

(But there's more.)

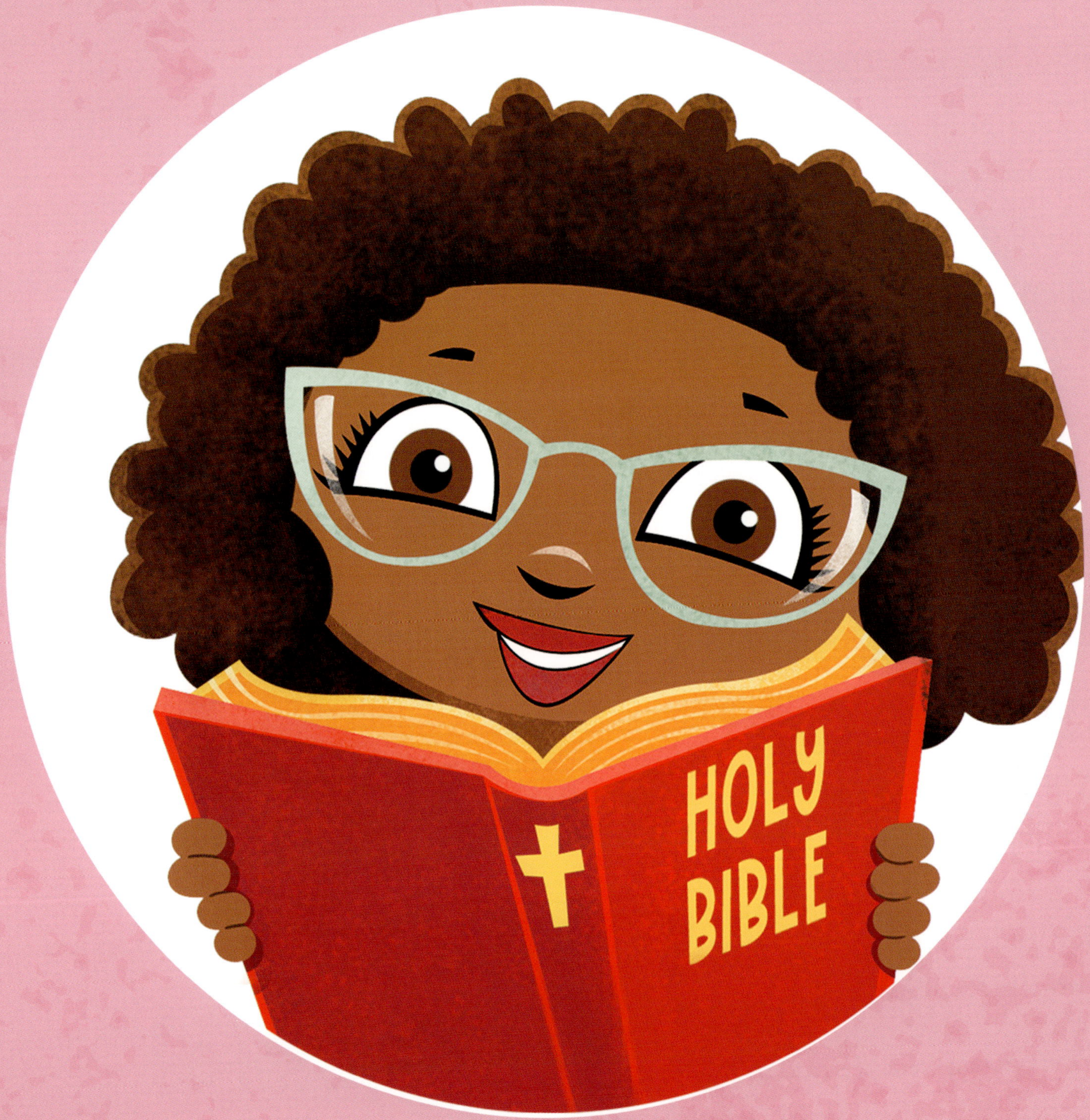
HOLY
BIBLE

What does the Bible say?

The Bible has **a lot** to say about how and why God created us. God designs everything with a purpose. Girls and boys are the same in some ways and very different in others. This makes life *wonderful*. Here are a few verses that can remind you how **special** you are.

1. "So God created mankind in His own image, in the image of God He created them; male and female he created them." **Genesis 1:27 (NLT)**

2. "For You made the parts inside me. You put me together inside my mother." **Psalms 139:13 (NLV)**

3. "For we are God's masterpiece. He has created us anew in Christ Jesus, so we can do the good things he planned for us long ago." **Eph 2:10 (NLT)**

4. "Don't copy the behavior and customs of this world, but let God transform you into a new person by changing the way you think..." **Romans 12:2 (NLT)**

5. "For this is how God loved the world: He gave His one and only Son, so that everyone who believes in him will not perish but have eternal life." **John 3:16 (NLT)**

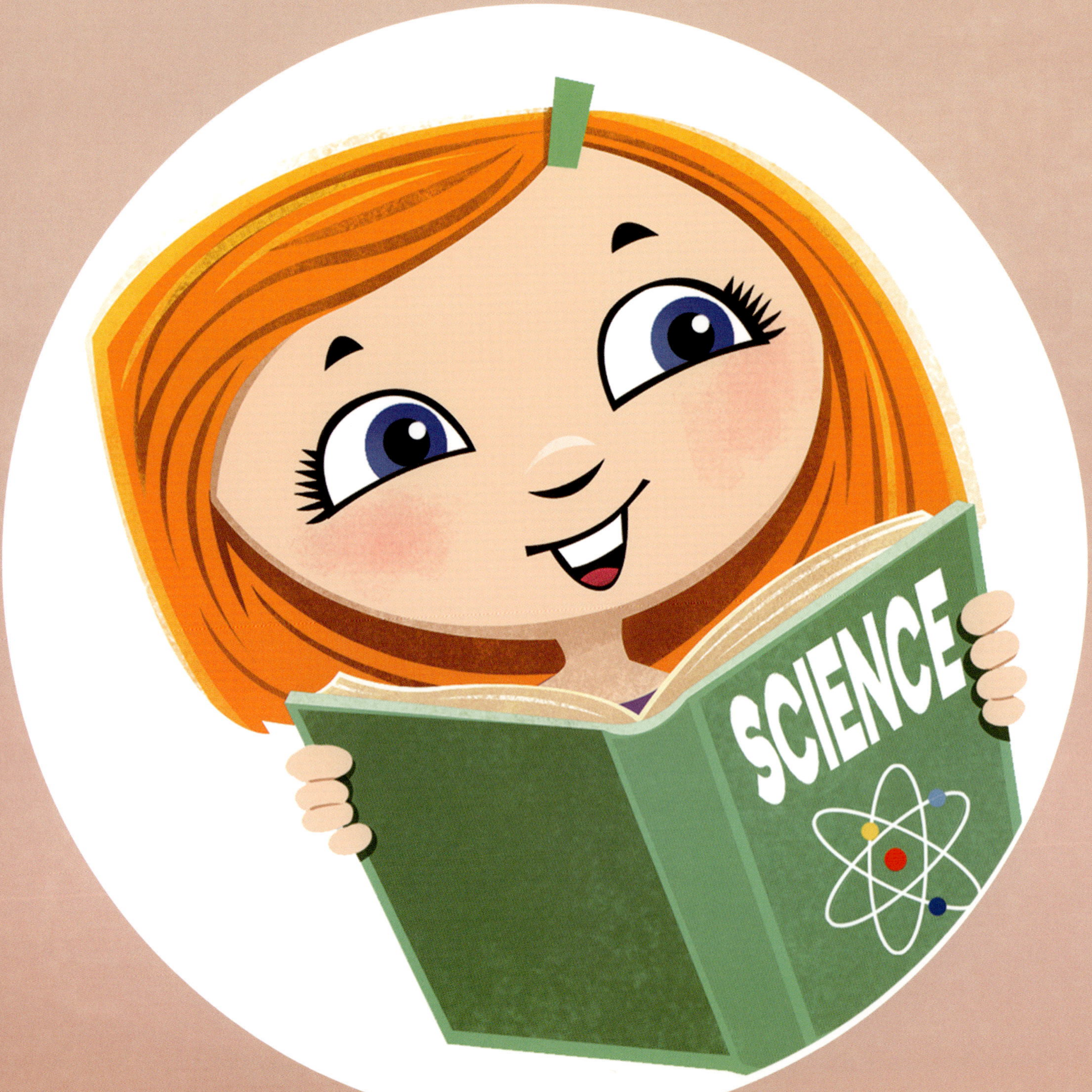
SCIENCE

What does science say?

God created the Laws of Nature, which means He created science. Science repeatedly proves Biblical truths correct over and over again. Here are some scientific facts about how girls and boys are different in important ways. It starts **way** before we're born!

1. As soon as we exist, something special inside each of us (called DNA) determines whether we will be girls or boys. That DNA never changes, no matter **how** we feel.

2. Doctors know **long** before we're born whether we are females or males. They use a special machine to see inside a mother's womb. The image is called an **ultrasound**.

3. There are **thousands** of physical differences between girls and boys. From our brains to our faces to our lungs and other body parts, we are wonderfully equal but not the same.

4. From running to swimming to soccer and volleyball, it's important to have separate girls' & boys' sports teams to give us **all** a chance to **shine**. It's fun to compete when it's fair.

5. Only females can get pregnant, carry babies inside of them and give birth. How neat! No guy can ever do that. Even so, moms and dads are equally valuable.

NOTE: Get more info and resources at **SHEiSSHE.COM**

about the authors

Bethany and Ryan Bomberger are the founders of The Radiance Foundation, a faith-based, educational, nonprofit organization. Their reach is global as they powerfully illuminate that every human life has God-given purpose. Their innovative and bold work has earned massive mainstream media coverage including the NY Times, MSNBC, Fox News, CNN, The Christian Post, ABC News, NewsMax, Breitbart, NPR, Epoch Times, World, Washington Times, Washington Post, NC Register, JET, LA Times, Yahoo! News and many more.

Bethany Bomberger is an educator by profession. She taught for over a decade in public and private schools in both suburban and urban settings. As a homeschooling mama, she's an advocate of school choice. Bethany is the Executive Director of The Radiance Foundation. She's an international public speaker, podcaster (LifeHasPurpose.com) and author of the ground-breaking children's book **PRO-LIFE KIDS!**. From women's conferences to pregnancy center galas to Supreme Court rallies, she loves speaking about how Christ makes us stronger than our circumstances.

Ryan Bomberger is the Chief Creative Officer of The Radiance Foundation. He's an international public speaker, columnist, factivist, and author of **Not Equal: Civil Rights Gone Wrong**. He has a passionate perspective on purpose. He was adopted and loved in a diverse family of fifteen (ten were adopted). Today, he's an Emmy Award-winning creative professional who designs messaging that's fearless, factual and freeing. Ryan has keynoted at Harvard, Princeton, Columbia Law School, Penn State, Howard University, Eureka College, Quinnipiac, and many more events around the world.

Bethany and Ryan met at Regent University where they earned their Master's degrees. They are the blessed parents of four awesome kiddos (two of whom were adopted): Rai Rai, Kai, Aliyah and Justice. Their hearts are to build a culture that values life at every age and stage.

Visit us at www.radiance.life!

about the illustrator

Ed Koehler is a freelance illustrator specializing in fun, lively art for children's books, educational materials, and any variety of print and online products. His work has been published around the world, and he's received numerous awards from the Evangelical Press Association and Associated Church Press. He's a member of the Society of Children's Book Writers and Illustrators, and the St. Louis Artists Guild. Working with publishers, designers, and product developers throughout the world, Ed creates fun art for books, curriculum, magazines, packaging and all kinds of kid-friendly products.

Bethany and Ryan first worked with Ed who illustrated the history-making **Pro-Life Kids!** book. People across the globe have been entertained and inspired by Ed's whimsical style. Together, the Bombergers and Koehler have created another life-affirming book in **She is She** and hope to shift culture through creativity, truth and love.

PLEASE VISIT US AT

SHE iS SHE.COM

When your identity is rooted in Christ, it won't be uprooted by everything else.

—Bethany & Ryan Bomberger